中国沼气工程案例图鉴——生物天然气
（第二辑）

郭建斌　刘秋琳　王凯军　李景明　董仁杰　著

清华大学出版社
北京

内容简介

本书系统介绍中国生物天然气工程建设情况，通过实地调研国内多个地区的生物天然气工程，选取了15个具有代表性的典型案例，从建设概况、厂区全貌、工艺路线、示范意义等方面，全方位展示了生物天然气工程全貌以及先进设计理念与技术创新。本书对我国生物天然气行业先进技术、典型模式与示范项目等进行了梳理与总结，可指导我国生物天然气工程建设，为相关从业人员和从事相关领域研究的科研人员以及环境科学与工程专业相关人士提供参考与借鉴。

版权所有，侵权必究。举报：010-62782989，beiqinquan@tup.tsinghua.edu.cn。

图书在版编目(CIP)数据

中国沼气工程案例图鉴. 第二辑, 生物天然气 / 郭建斌等著. -- 北京：清华大学出版社, 2024.10.
ISBN 978-7-302-67480-1

Ⅰ. S216.4-64
中国国家版本馆CIP数据核字第2024TC0254号

责任编辑：张占奎
封面设计：沐风堂
责任校对：王淑云
责任印制：丛怀宇

出版发行：清华大学出版社
网　　址：https://www.tup.com.cn, https://www.wqxuetang.com
地　　址：北京清华大学学研大厦A座　　　邮　编：100084
社 总 机：010-83470000　　　　　　　　　邮　购：010-62786544
投稿与读者服务：010-62776969, c-service@tup.tsinghua.edu.cn
质量反馈：010-62772015, zhiliang@tup.tsinghua.edu.cn

印 装 者：北京博海升彩色印刷有限公司
经　　销：全国新华书店
开　　本：285mm×260mm　　　印　张：13$\frac{1}{3}$　　　字　数：132千字
版　　次：2024年10月第1版　　　　　　　　印　次：2024年10月第1次印刷
定　　价：198.00元

产品编号：109516-01

作者简介
Author Profile

郭建斌

中国农业大学工学院副教授，博士生导师。科技部生物质能科学与技术国际联合研究中心副主任、中国沼气学会双碳专委会副主任。《农业机械学报》青年编委，联合国粮农组织畜牧业环境评估及绩效伙伴关系技术指导组成员，国际标准化组织 ISO/TC 255 专家组成员。长期致力于城乡有机废弃物的能源化、资源化利用。主持或参与国家、省部级等项目/课题 10 余项，发表国内外期刊论文 80 余篇。获省部级奖励 3 项，学会奖励 2 项。

刘秋琳

清华大学环境学院环境工程硕士，现就职于清华大学环境学院。中国沼气学会副秘书长，国家环境保护技术管理与评估工程技术中心主任助理，境界平台联合创始人，长期致力于环境领域技术推广与产业研究。参与水体污染控制与治理科技重大专项等国家课题；参编《污染防治可行技术指南编制导则》等国家环境保护标准；著有《中国沼气工程案例图鉴（第一辑）》《城市污泥干化焚烧工程实践》，参编《环保回忆录》《地下再生水厂览胜》《北方大型人工湿地工法与营造》等。

王凯军

清华大学环境学院教授，历任中国沼气学会秘书长、理事长，国家环境保护部科学技术委员会委员，国家环境保护技术管理与评估工程技术中心主任，国家水体污染控制与治理科技重大专项总体组专家。在荷兰 Wageningen 农业大学环境技术系获得博士学位，师从国际厌氧大师 Lettinga 教授。主要研究方向：城市污水和工业废水处理与资源化理论及方法，城市和农业废弃物处理与可再生能源技术开发，环境保护政策、标准研究与产业化方向。出版《厌氧生物技术（Ⅰ）理论与应用》《厌氧生物技术（Ⅱ）工程与实践》等厌氧领域多本专著。

李景明

研究员,毕业于清华大学工程力学系。长期从事农村能源和环境保护科研开发、行政管理和技术推广工作。现任中国沼气学会秘书长、《中国沼气》副主编、全国沼气标准化技术委员会秘书长。曾先后组织制修订国家和行业标准 100 多项,主持国家和省部级科研、技术推广和国际合作项目 20 余项;曾获国家科技进步奖二等奖 1 项,国家能源科技进步奖三等奖 1 项;主编书籍 7 本,发表论文 50 余篇。

董仁杰

中国农业大学工学院教授,科技部生物质能科学与技术国际联合研究中心主任,农业农村部可再生能源清洁化利用重点实验室常务副主任,国际标准化组织沼气工程标委会主席,国际能源署生物能源沼气工程工作组中国代表。2024 年度全球前 2% 顶尖科学家。致力于生物质能开发与温室气体减排 35 年,主张"应气尽气"和"光电转气",通过减污降碳、能源脱碳和自然固碳创建农业农村绿色发展新模式。发表论文 150 余篇。曾获国家科技进步奖二等奖、神农中华农业科技奖二等奖。

序
Preface

从 2015 年起，我国沼气行业开始转型升级，国家政策逐步转向支持大型、工业化、规模化沼气及生物天然气项目。2019 年，由国家发改委和农业农村部等 10 个部委联合印发《关于促进生物天然气产业化发展的指导意见》（发改能源规〔2019〕1895 号），提出到 2030 年，生物天然气实现稳步发展，规模位居世界前列，生物天然气年产量超过 200 亿 m^3，占国内天然气产量一定比重。2023 年，在全面深化改革的推动下，中央又提出了能耗双控逐步转向碳排放双控的政策，为生物天然气这一具有负碳排放特性的绿色能源提供了新的发展空间。

随着"双碳"战略的推进，生物天然气将成为未来中国绿色能源版图中的重要组成部分。当前我国生物天然气生产从预处理、厌氧工艺、沼气净化提纯以及沼液沼渣综合利用等方面基本达到可以依据原料特性、产业特点，形成与行业政策相符的发展模式，初步实现了废弃生物质资源的肥料化和能源化利用，涌现了一大批技术先进、运行良好的工程示范项目，在可再生能源利用与温室气体减排方面发挥了重要作用。

2023 年，中国沼气学会组织编写了《中国沼气工程案例图鉴（第一辑）》。在此基础上，2024 年中国沼气学会又组织相关人员，聚焦农业、城市、工业生物天然气工程项目，进行了密集走访与调研，并从中遴选了若干优秀生物天然气项目案例，制成工程图鉴，拟向读者呈现我国生物天然气产业发展的新面貌。在本书编写过程中，得到了多家沼气和生物天然气企业和多位行业同仁的支持，并为本书贡献了工程案例及图文素材，在此表示感谢。

限于时间、资料和作者水平，不足之处在所难免，欢迎业界朋友提出宝贵意见和建议，共同为我国顺利实现"双碳"目标和沼气与生物天然气事业的健康发展而努力！

作者
2024 年 8 月 5 日

目录
Contents

001 洛碛厨余垃圾资源化利用项目	**084** 射阳规模化生物天然气项目
014 应县畜禽粪污资源化整县推进项目	**094** 大丰畜禽粪污集中处理综合利用项目
026 三河天龙新型建材有限公司规模化生物天然气试点工程	**104** 镇江市餐厨废弃物及生活污泥协同处理项目
038 故城福隆规模化生物天然气项目	**114** 佛山可再生能源（沼气）制氢加氢母站项目
046 济南市餐厨垃圾无害化处理与资源化利用项目	**122** 大湾区餐厨废弃物制备生物天然气示范项目
058 诸城舜沃大型沼气提纯项目	**134** 造纸废水沼气提纯制备生物天然气示范项目
070 商水生物天然气项目	**150** 致谢

洛碛厨余垃圾资源化利用项目

该项目位于重庆渝北区洛碛镇桂湾村，由重庆市环卫集团建设运营，总投资约22.4亿元，占地约393亩（1亩=666.67m²），是全国多源有机固废单体处置规模最大的厨余垃圾资源化利用项目。该项目设计规模年处理餐厨垃圾76.65万t、厨余垃圾36.5万t，主要采用干式、湿式厌氧消化工艺，实现餐厨垃圾、厨余垃圾、污泥联合厌氧消化生产新能源，年产生沼气9,400万m³，发电量1.3亿kW·h，天然气350万m³，餐厨废弃油脂3.5万t。

Luoqi kitchen waste resource utilization project

Luoqi kitchen waste resource utilization project is located in Yubei District, Chongqing. It is the largest food, kitchen waste and sludge resource utilization project in China. The plant could treat 766,500 t food waste, 365,000 t kitchen waste, with an annual biogas production 94 million m³, power generation of 130 million kW·h, 3.5 million m³ bio-natural gas and 35,000 t kitchen waste oil.

工艺路线

餐厨垃圾处理工艺主要包括分拣、制浆、除砂除杂、提油、湿式厌氧罐、沼液处理等单元。

厨余垃圾处理工艺主要包括破碎、筛分、中温干式厌氧罐、沼液处理等单元。

厌氧发酵产生的沼气经过脱硫、脱碳净化后用来发电以及制备天然气。

沼液通过两级格栅+高效组合气浮+Bardenpho生化脱氮除磷工艺+Fenton高级氧化+曝气生物滤池（biological aerated filter，BAF）技术等工艺处置后进行污水达标排放。

Process route

The food treatment process mainly includes sorting, pulping, sand and impurity removal, oil extraction, wet anaerobic and digestate treatment.

The kitchen waste treatment process mainly includes crushing, screening, dry anaerobic tank and digestate treatment.

Biogas is desulfurized, decarbonized for power generation and bio-nature gas production.

Liquid digestate is treated by the two-stage grid + high-efficiency combined air flotation +Bardenpho biochemical nitrogen and phosphorus removal process +Fenton advanced oxidation +BAF technology to meet the standard.

工艺流程图
Process route chart

餐厨垃圾预处理单元

9 条分拣制浆生产线，16 条三相离心提油生产线；
日处理餐厨垃圾 2,100 t，回收动植物油脂约 100 t。

Food waste pretreatment unit

Nine sorting and pulping production lines and 16 three-phase centrifugal oil extraction production lines;
2,100 t food waste daily, and about 100 t waste oils recovered.

厨余垃圾预处理单元

设有 2 条生产线，生产线包括破碎、磁选、筛分等主要设备，日处理厨余垃圾 500 t。

Kitchen waste pretreatment unit

There are two production lines, including crushing, magnetic separation, screening etc, daily processing capacity of 500 t kitchen waste.

∨ 餐厨垃圾预处理单元
Food waste pretreatment unit

∨ 厨余垃圾预处理单元
Kitchen waste pretreatment unit

湿式厌氧发酵单元

6 个湿式厌氧发酵罐，单个有效容积 12,000 m^3；

8 条离心脱水处置线，日处理餐厨垃圾 2,100 t，日产沼气约 18 万 m^3。

Wet anaerobic digestion unit

6 wet AD tanks, 12,000 m^3 of each.

8 centrifugal dewatering disposal lines, daily processing 2,100 t restaurant garbage and producing about 180,000 m^3 biogas.

∨ 湿式厌氧发酵单元
Wet anaerobic digestion unit

干式厌氧发酵单元

7 个卧式干式厌氧发酵罐,总容积 14,550 m³;3 个竖式干式厌氧发酵罐,总容积 8,850 m³;日处理厨余垃圾 1,000 t,日产沼气约 10 万 m³。

Dry anaerobic digestion unit

7 horizontal dry AD tanks with total volume 14,550 m³. 3 vertical anaerobic fermentation tanks with total volume 8,850 m³; daily processing 1,000 t kitchen waste and producing about 100,000 m³ biogas.

∨ 干式厌氧发酵单元
Dry anaerobic digestion unit

沼气净化单元

净化工艺为生化湿法脱硫 + 干法脱硫；

日处理沼气 27 万 m³，H_2S 含量由不高于 0.2% 处理至不高于 0.005%。

Biogas desulfurization unit

The purification process is biochemical wet desulfurization+dry desulfurization.

The daily treatment of biogas 270,000 m³, and the H_2S content is reduced from 0.2% to 0.005%.

∨ 沼气净化单元
Biogas desulfurization unit

沼气提纯单元

提纯工艺为沼气胺法脱碳；
经处理后的沼气甲烷浓度达 97% 以上，甲烷回收率超过 98%。

Biogas purification unit

Amine method is applied to produce biomethane.
The bio-natural gas concentration can reach over 97%, and recovery rate is over 98%.

沼气提纯单元 >
Biogas purification unit

发电系统

设有 12 台发电机组，发电能力 1.8 MW·h，并配有 4 套余热锅炉回收热能。

日处理沼气 18 万 m^3，发电约 36 万 kW·h，在保障厂区用电基础上将剩余电力并入市政电网。

Power generation system

12 generator, power generation capacity of 1.8 MW·h, and 4 sets of waste heat boilers to recover heat energy.

Daily biogas consumption 180,000 m^3, 360,000 kW·h electricity generation, for the factory running and municipal power grid injection.

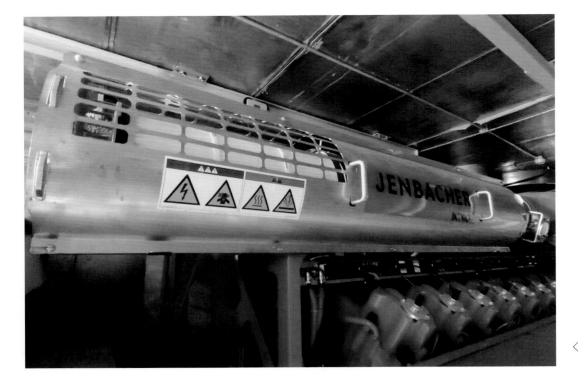

〈 发电系统
Power generation system

锅炉单元

2台6t锅炉，3台20t锅炉。
采用低N燃烧技术，烟气达到重庆市地方排放标准。

Boiler unit

Two 6-ton boilers and three 20-ton boilers.
Using low N combustion technology, the flue gas reaches the local emission standard of Chongqing.

锅炉系统
Boiler unit

洛碛厨余垃圾资源化利用项目

示范意义

建立了餐厨、厨余垃圾统一收运、集中处理、多源有机固废协同处置模式，构建了"垃圾—资源—产品"再生资源循环完整产业链。该项目每年可代替标煤5.3万t，减少CO_2排放13.3万t，减少NO_x排放2,000 t，减少SO_2排放4,000 t。

Demonstration significance

The Luoqi Project has established a unified collection, transportation, centralized treatment and multi-source organic solid waste collaborative disposal mode in the main urban area of Chongqing, which realized the complete industrial chain of "waste-resource-product". The project can replace 53,000 t standard coal every year, reducing CO_2 emissions by 133,000 t, NO_x emissions by 2,000 t and SO_2 emissions by 4,000 t.

洛碛厨余垃圾资源化利用项目

应县畜禽粪污资源化整县推进项目

该项目由山西华新燃气集团建设，于2022年建成，占地121亩，一期投资1.36亿元，采用"原料预处理+中温厌氧发酵+三沼综合利用"技术路线，可年处理牛粪20余万t，年产沼气近1,100万 m^3、生物天然气460万 m^3、沼渣6万t、沼液11.7万t，减排温室气体8.5万 t/a。

Ying County animal manure resource utilization project

The project is completed in 2022. Annually, the plant treats more than 200,000 t cow manure through mesophilic AD. Nearly 11 million m^3 biogas, 4.6 million m^3 bio-natural gas, GHG emission reduction reach 85,000 t per year.

应县畜禽粪污资源化整县推进项目

工艺路线

采用预处理、全混式厌氧发酵工艺，沼气经脱硫提纯后所得到的生物天然气用于陶瓷产业，沼液用于盐碱地改良和灌溉，沼渣回用于牧场作为垫料。

Process route

The pretreatment + mesophilic AD + products comprehensive utilization is used. Bio-natural gas is injected into the local gas pipeline network, which is used in the Ying County ceramic industry cluster. The liquid digestate is used for saline land improvement and irrigation, and the solid digestate is reused as bedding material.

▽ 项目工艺流程图
Process flow chart of the project

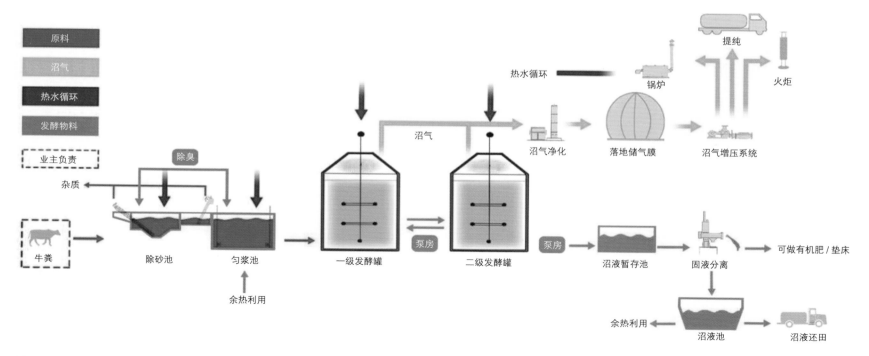

应县畜禽粪污资源化整县推进项目

预处理单元

牛粪除砂、匀浆工艺。

Pretreat unit

Desanding and homonization.

∨ 匀浆池
Homogenizer

∨ 除砂装置
Desanding device

应县畜禽粪污资源化整县推进项目

厌氧消化单元

项目建厌氧发酵罐 4 座（单体有效容积 6,000 m³），采用两组两级 CSTR 中温厌氧发酵工艺。

Anaerobic digestion unit

Four anaerobic fermenters (effective volume of 6,000 m³ per unit) are built, using two groups of two-stage CSTR medium temperature anaerobic fermentation process.

沼气净化提纯单元

沼气通过脱硫提纯后，产生的生物天然气并入燃气管网，直接运用于应县陶瓷产业集群。

Biogas desulfurization and purification unit

The bio-natural gas is incorporated into the gas pipeline network after desulfurization for the Ying County ceramic industry.

沼气净化与提纯设备
Biogas desulfurization and purification equipment

沼液沼渣利用单元

沼液用于灌溉还田、盐碱地修复、微藻培养固碳。

沼渣用于牛床垫料。

Digestate utilization unit

Liquid digestate is used for saline and alkaline land improvement, irrigation and microalgae cultivation.

Solid digestate is used for bedding material.

∨ 微藻固碳
Microalgae cultivation

∨ 沼液还田
Slurry to the farmland

应县畜禽粪污资源化整县推进项目

沼渣垫料
Solid digestate bedding

应县畜禽粪污资源化整县推进项目

示范意义

本项目为山西省生物能源环保产业的示范工程，致力于实现生物质能源与陶瓷产业、生态农业的耦合发展，提高当地畜禽养殖粪污资源化利用水平，也为当地工业燃料的提档升级和低碳化改造提供了有力支撑，起到了很好的示范带动作用。

Demonstration significance

This project is committed to realizing the coupling development of biomass energy, ceramic industry and ecological agriculture, providing strong support for the upgrading and low-carbon transformation of local industrial fuel.

三河天龙新型建材有限公司规模化生物天然气试点工程

该项目位于河北省廊坊市,由三河市盈盛生物能源科技股份有限公司投资,占地109亩,总投资2亿多元,北京盈和瑞环境科技有限公司负责EPC总承包,采用"纤维素水解预处理技术+厌氧发酵"工艺处理农业固体废弃物,实现规模化生产后年可消纳农作物秸秆11万t、畜禽粪污等其他有机废弃物20万t,年减排CO_2 10万t。

Sanhe Tianlong large-scale bio-natural gas project

The project is located in Langfang, Hebei. It is invested by Sanhe Ying Shen Biology Energy Technology Co., LTD, covering 109 acres with a total investment of more than 200 million yuan. Beijing Ying He Rui Environmental Technology Co., LTD is responsible for the EPC, and adopts the "semi-aerobic hydrolysis + anaerobic fermentation" processes to treat organic waste. It can annually process 200,000t organic waste and 110,000 crop straw, reducing carbon dioxide emissions 100,000 t.

三河天龙新型建材有限公司规模化生物天然气试点工程

工艺路线

采用"秸秆兼氧纤维素水解预处理 + 两级除砂 +CSTR 厌氧发酵 + 生物脱硫 + 三级膜提纯"的工艺路线。

Procss route

Semi-aerobic hydrolysis + two-stage sand removal + CSTR anaerobic fermentation + biological desulfurization + three-stage membrane purification.

∨ 项目工艺流程图
Process flow chart of the project

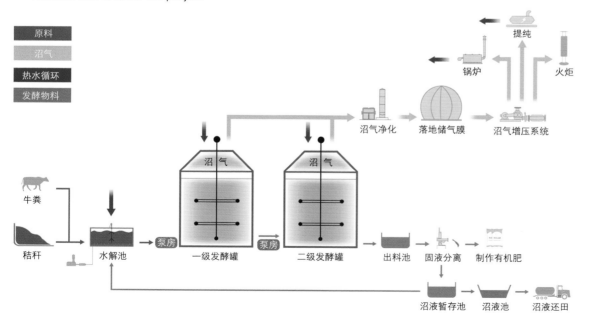

三河天龙新型建材有限公司规模化生物天然气试点工程

预处理单元

采用纤维素水解预处理技术降解有机物质。

控制温度、含水等参数，促进微生物的生长和代谢。

Pretreatment unit

Cellulose hydrolysis pretreatment technology.

Temperature and water are controlled to promote the growth of microorganisms.

∨ 水解池
Hydrolysis tank

∨ 进料箱
Feed box

三河天龙新型建材有限公司规模化生物天然气试点工程

厌氧发酵单元

厌氧发酵罐：3,600 m³×9 座，两级 CSTR 厌氧工艺。

Anaerobic digestion unit

Nine tanks with each volume 3,600 m³.
Two-stage CSTR AD process.

∨ 厌氧发酵区
Anaerobic digestion zone

沼气净化提纯单元

提纯单元由脱硫、预增压系统、膜分离系统、增压系统和充装系统组成。

Biogas desulfurization and purification unit

It consists of desulfurization, pre-pressurization, membrane separation, pressurization and charging system.

∨ 沼气脱硫系统
Biogas desulfurization system

三河天龙新型建材有限公司规模化生物天然气试点工程

生物甲烷指标：CH$_4$ 不低于 97%，或者满足一类天然气标准；

甲烷回收率：不低于 99%；

单位能耗：0.20~0.25 kW·h/Nm^3 沼气；

运行压力：中压，10~20 bar；

生物甲烷利用方式：CNG 或者燃气管网。

Methane content: Not less than 97% or meeting first-class natural gas standards.

Methane recovery rate: Not less than 99%.

Specific energy consumption: 0.20~0.25 kW·h/Nm^3 biogas.

Operating pressure: Medium pressure, 10~20 bar.

Bio-natural gas utilization: CNG or gas grid.

∨ 沼气提纯系统与加气站
Biogas purification system and gas station

三沼利用单元

沼气利用：沼气经三级膜提纯工艺深度加工为绿色生物天然气。生物天然气可用于发电、供暖、烹饪等。

沼渣利用：沼渣用作土壤改良剂和肥料施用于农田。

沼液利用：沼液用于农田灌溉或者与沼渣混合后喷施于农田。

Biogas & digestate utilization unit

Biogas utilization: Bio-natural gas is produced by three-stage membrane purification process. It can be used for power generation, heating, cooking and other purposes.

Solid digestate utilization: It can be used as a soil conditioner and fertilizer for farmland.

Liquid digestate utilization: It is used for farmland irrigation.

沼气提纯为生物天然气
Biogas is purified into bio-natural gas

∨ 固体有机肥利用
Solid organic fertilizer utilization

∨ 液体有机肥利用
Liquid organic fertilizer utilization

示范意义

项目年产沼气 1,200 万 m^3、生物天然气 530 万 m^3，有效解决了三河市域内及京津周边范围内玉米、小麦秸秆农作物副产品、畜禽粪便等有机废弃物的去向问题，有利于减少环境污染，提升农业废弃物的综合回收能力，消除安全隐患。

Demonstration significance

With an annual output of 12 million m^3 biogas and 5.3 million m^3 bio-natural gas, the project effectively solves the problem of organic waste such as corn, wheat straw, crop by-products, livestock and poultry manure in Sanhe City, reduces environmental pollution, improves the nutrient recycling of agricultural waste.

三河天龙新型建材有限公司规模化生物天然气试点工程

故城福隆规模化生物天然气项目

该项目位于河北省衡水市,由杭州能源环境工程有限公司建设,于2017年建成,占地136亩,总投资9,000万元,采用"粪污预处理+厌氧发酵+沼气净化提纯+固液分离和有机肥生产"的技术路线,每日可处理猪粪316 t、猪场养殖废水450 t及玉米秸秆21 t,可日产生物天然气1.5万 m^3、沼液689 t、沼渣66 t,年减排$CO_2$7.5万 t。

Gucheng Fulong large-scale bio-natural gas project

The project is located in Hengshui City, Hebei Province, built by Hangzhou Energy & Environmental Engineering Co., Ltd in 2017. Adopting the technological route of manure pretreatment + anaerobic digestion + biogas purification + solid - liquid separation and organic fertilizer production. Capable of processing 316 t pig manue, 450 t wastervater from pig farms, and 21 t corn stalks per day. It can produce 15,000 m^3 of bio-natural gas per day and reduce CO_2 emissions by 75,000 t annually.

故城福隆规模化生物天然气项目

2015 年国家发改委、农业部生物天然气试点项目
2015 National Development and Reform Commission (NDRC) and Ministry of Agriculture (MOA) Bio-nature Gas Project

工艺路线

采用"粪污预处理 + 厌氧发酵 + 沼气净化提纯 + 固液分离和有机肥生产"的工艺路线。

Process route

The technical process is: manure pretreatment + anaerobic digestion + biogas purification + solid-liquid separation and organic fertilizer production.

∨ 工艺流程图
Process flow chart

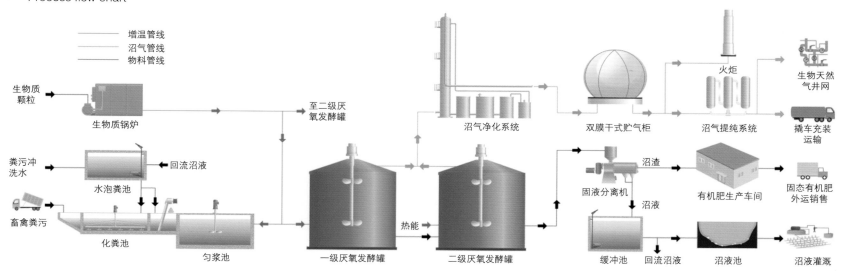

预处理单元

通过化粪池、调节池与匀浆池对原料进行均质与预增温。

Pretreatment unit

Through the septic tank, regulating tank and homogenizing tank, the raw material is homogenized and pre-heated.

∨ 预处理单元
Pretreatment unit

厌氧发酵单元

4 座 6,000 m³ 的 CSTR 厌氧发酵罐,采用中温厌氧发酵工艺,物料停留时间大于 30 天。

Anaerobic digestion unit

Four 6,000 m³ CSTR mesophilic anaerobic digestion rectors, material retention time >30 days.

∨ 厌氧发酵单元
Anaerobic digestion unit

沼气净化提纯单元

采用生物脱硫工艺，处理后沼气中硫化氢含量不大于 15 mg/m³。此外，采用膜分离技术对脱硫后的沼气进行脱碳提纯。

Biogas desulfurization and purification unit

The hydrogen sulfide content in biogas after treatment is no more than 15 mg/m³ by biological desulfurization process. Membrane separation was used to decarbonize and purify the desulfurized biogas.

▽ 沼气净化单元
Biogas desulfurization unit

▽ 提纯单元
Purification unit

示范意义

结合故城县规模化养殖企业粪污处理要求，猪粪通过集中处理转化为清洁能源和有机肥，实现种养结合，极大地减轻了地方环保生态压力，也为当地能源供给提供了有力保障，是地方政府在生物天然气领域跨出的重要一步。

Demonstration significance

In combination with the manure treatment requirements of large-scale breeding enterprises in Gucheng County, the pig manure is transformed into clean energy and organic fertilizer through centralized treatment, which is an important step for the local government in the field of bio-natural gas.

故城福隆规模化生物天然气项目

济南市餐厨垃圾无害化处理与资源化利用项目

项目由山高环能集团股份有限公司下属公司建设，位于济南市起步区，占地面积 30.23 亩，总投资 2.4 亿元，特许经营期 25 年，餐厨、厨余垃圾日处理规模 480 t。该项目收运、处置一体化，覆盖济南九个区县的餐厨、厨余垃圾处理。将餐厨、厨余垃圾经数字化收运进厂，并对其进行无害化处理和资源化利用，年处理餐厨、厨余垃圾 16 万 t，生物质天然气生产能力 365 万 m^3/ 年，年减排量 3.92 万 t CO_2e。

Jinan City food waste harmless treatment and resource utilization project

The project is constructed by a subsidiary company of Shangao Environmental Energy Group Co., LTD. As one of the third batch of national pilot city demonstration projects, the plant is located in Jinan City, covering an area of 30.23 mu. 240 million yuan was invested with the franchise period of 25 years. Daily processing capacity of food and kitchen waste is 480 t. The project covers the food waste disposal of nine districts and counties in Jinan. The annual treatment of food waste is 160,000 t, the annual production capacity of bio-natural gas is 3.65 million m^3, and the annual emission reduction is 39,200 t CO_2e.

济南市餐厨垃圾无害化处理与资源化利用项目

工艺路线

餐厨、厨余废弃物处理工艺主要包括智能化收运、分选制浆、浆液提油、厌氧发酵、生物天然气生产、沼液处理等单元。

Process route

The food and kitchen waste treatment process mainly includes intelligent collection and transportation, separation and pulping, oil extraction, anaerobic fermentation, bio-natural gas production, and digestate treatment.

项目工艺流程图
Process flow chart of the project

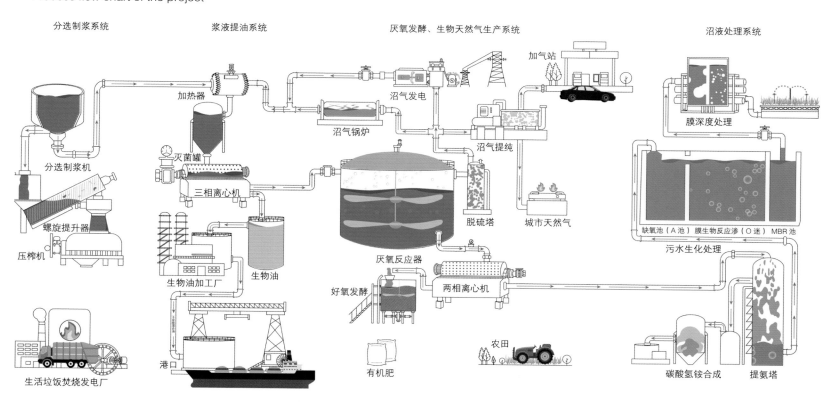

预处理单元

采用"水力制浆 + 除杂除砂 + 浆液提油"的预处理工艺,可将固渣含水率降至 65% 以下,出渣率小于 15%,提油率较常规破碎工艺提高 0.5%~1%。

Pretreatment unit

The hydraulic pulping + impurity and sand removal + wet heat oil extraction is used under pretreatment system. The process can reduce the water content of solid to less than 65%, the solid yield is less than 15%, and the oil extraction rate is 0.5%~1% higher than the conventional process.

∨ 水力旋流制浆
Hydrocyclone pulping

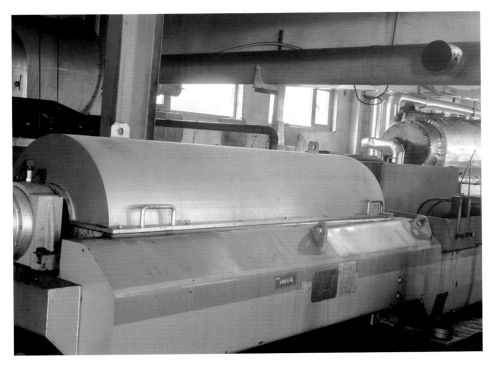

∨ 三相离心机
Three-phase centrifuge

厌氧发酵单元

采用湿式中温双环式两相厌氧发酵工艺，内外环总容积约 5,800 m³，产酸产甲烷在外环和内环中进行。

Anaerobic digestion unit

The mesophilic wet double-loop two-phase AD process is used. The total volume of the inner and outer rings is about 5,800 m³. The acid and methane production is carried out in the outerring.

∨ 双环式两相厌氧发酵
Double-loop two-phase anaerobic fermentation

污水处理单元

采用"汽提脱氨+A/O+MBR+NF+RO+物料膜+浓液处理"工艺处理污水。

Sewage treatment unit

The ammonia stripping +A/O +MBR +NF +RO +material film+ concentrated liquid treatment is used in sewage treatment unit.

污水处理单元 >
Sewage treatment unit

沼气净化单元

采用原位脱硫+干法脱硫的沼气净化工艺，日处理沼气 30,000 m³，H_2S 经厌氧系统原位脱硫后含量低于 0.1%，再经干法脱硫处理至低于 0.001%。

Biogas desulfurization unit

In-situ desulfurization + dry desulfurization, the daily treatment of biogas is 30,000 m³, H_2S content after in-situ desulfurization is less than 0.1%, and then dry desulfurization treatment to less than 0.001%.

> 脱硫塔
Desulfurization tower

沼气提纯单元

采用膜分离技术，沼气经脱硫后进入提纯净化系统，经加压、脱水、脱除 VOC 后进行膜分离脱碳。产品天然气 CH_4 回收率可达 97% 以上。

Biogas purification unit

After pressurization, dehydration and VOC removal, membrane separation technology is applied for bio-natural gas production. The recovery rate of bio-natural gas is higher than 97%.

∨ 膜系统
Membrane system

生物天然气利用单元

生物天然气产品可加压后通过车载瓶组运输销售,也可减压后纳入市政管网,还可以供厂内收运车辆(加气)使用。

Bio-natural gas utilization unit

The bio-natural gas can be sold as CNG, injected to the municipal pipe network, or supplied to the gas station.

∨ 天然气储气瓶组
Natural gas storage cylinder group

∨ 天然气纳管前减压撬
Decompression skid before natural gas piping

∨ 天然气加气枪
Gas gun

示范意义

本项目每年可处理餐厨、厨余废弃物 16 万 t，生成沼气、生物天然气、腐殖酸有机肥、碳酸氢铵肥、生物柴油原料等绿色资源化产品，有效防止"地沟油"等非法食品原料回流，保障食品安全，推动城市垃圾分类、无害化处理与资源化利用，提升城市环境治理和资源管理水平；助力打造绿色、低碳、可持续的城市环境。

Demonstration significance

160,000 t food and kitchen waste is treated annually, generating green products such as biogas, bio-natural gas, organic fertilizer, ammonium bicarbonate fertilizer, biodiesel raw materials.

济南市餐厨垃圾无害化处理与资源化利用项目

诸城舜沃大型沼气提纯项目

该项目位于山东省潍坊市,由青岛汇君环境能源工程有限公司建设,总投资约 9,000 万元。年处理养殖场粪便 59,787 t,秸秆 34,675 t,年产沼气 900 万 m³,进入天然气管网约 420 万 m³,年产沼渣肥 4.50 万 t,沼液肥 4.03 万 t。每年可代替标煤 4,749 t,减少 CO_2 排放 1.99 万 t,NO_x 排放 199.92 t,减少 SO_2 排放 352.8 t。

Zhucheng Shunwo large-scale bio-natural gas plant

The project is located in Weifang City, Shandong Province, constructed by Qingdao Conminent Environmental Energy Engineering Co., Ltd., with a total investment of about 90 million. yuan. The project treats 59,787 t manure and 34,675 t straw annually, with an annual bio-nature gas of 4.2 million m³. The project can reduce CO_2 emissions by 19,900 t, NO_x emissions by 199.92 t, and SO_2 emissions by 352.8 t.

诸城舜沃大型沼气提纯项目

工艺路线

以畜禽粪污和秸秆为原料，采用CSTR，两级湿式厌氧发酵、沼气提纯等技术，将有机废弃物经厌氧消化转化为沼气、沼渣、沼液。

Process route

Organic waste was converted into biogas, and digestate in two-stage CSTR. And biogas was purified into bio-natural gas.

∨ 项目工艺流程图
Process flow chart of the project

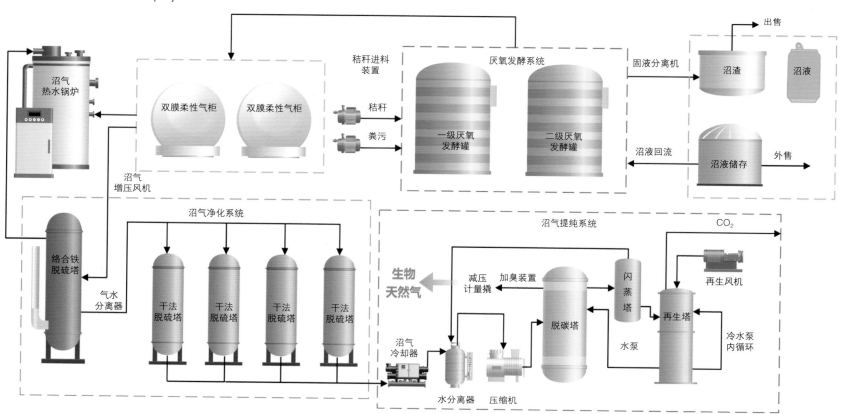

诸城舜沃大型沼气提纯项目

厌氧发酵单元

一级厌氧发酵罐（2个，5,000 m^3/个）+ 二级厌氧发酵罐（2个，5,000 m^3/个）+ 双膜气柜（2个，2,000 m^3/个）。
CSTR厌氧发酵罐内设4台潜水搅拌机和2台破壳搅拌机。

Anaerobic digestion unit

Two Primary (5,000 m^3 of each), two secondary CSTR (5,000 m^3 of each) and two double membrane gas storage (2,000 m^3 of each) are used.
CSTR is equipped with 4 diving mixers and 2 shell blender.

< 厌氧发酵单元
Anaerobic digestion unit

诸城舜沃大型沼气提纯项目

∨ 双膜气柜
Biogas storge

好氧发酵制肥单元

采用翻抛发酵工艺，发酵槽 2 个，尺寸为 10 m×90 m，年产有机肥 3 万 t。

Aerobic fermentation unit

Flip–throwing fermentation process is used with two aerobic fermentation slot size of 10 m×90 m. Organic fertilizer production capacity is 30,000 t/a.

∨ 有机肥生产车间
Organic fertilizer production workshop

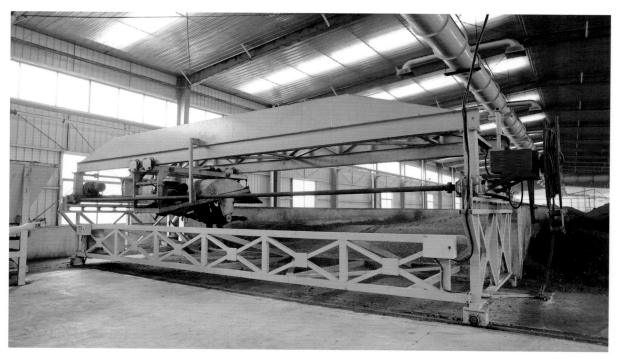

沼气净化单元

脱硫工艺为络合铁湿法脱硫 + 干法脱硫。

日处理沼气 22,000 m³，H_2S 含量由不高于 1% 处理至不高于 0.002%。

Biogas desulfurization unit

Combination of iron wet and dry desulfurization.

Daily processing biogas 22,000 m³, H_2S decreased from no more than 1% to no more than 0.002%.

∨ 沼气净化单元
Biogas desulfurization unit

沼气提纯单元

压力水洗提纯工艺。

合格的 BNG 产品达到一类天然气标准送至新奥燃气管网，CH_4 回收率为 98%±1%。

Biogas purification unit

Stress water scrubbing process is used.

Products are delivered to ENN gas network, and the CH_4 recovery rate is 98%±1%.

∨ 沼气提纯以及应用场景
Biogas purification and application scenarios

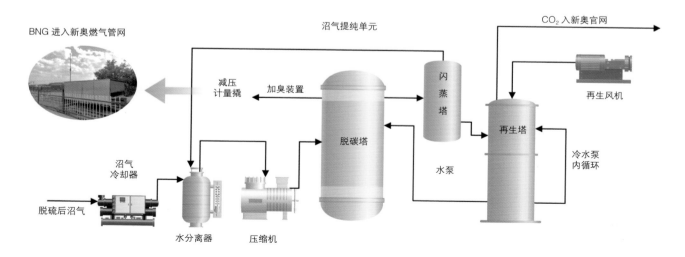

诸城舜沃大型沼气提纯项目

沼气提纯单元 〉
Biogas purification unit

示范意义

项目建立了区域有机废弃物第三方集中处理中心,防止畜禽粪便和农作物秸秆对地下水、土壤、空气的严重污染,生态效益显著。

建立了生物天然气多元化消费体系。开拓生物天然气在城镇居民炊事取暖、城市管网燃气、发电、交通燃料、锅炉燃料、工业原料等领域的应用。

Demonstration significance

The project has established a regional organic waste third-party centralized treatment center.

The bio-natural gas has been supplied for cooking and heating, pipeline injection, power generation, transportation fuel, boiler fuel, and industrial raw materials.

诸城舜沃大型沼气提纯项目

商水生物天然气项目

该项目由必奥新能源科技有限公司建设，2024年7月通过验收。项目采用必奥公司独家拥有的国际先进塞流式干式厌氧发酵技术，年处理玉米秸秆18万~20万t，可年产生物天然气1,878万 Nm^3，食品级二氧化碳2.2万t，固体生物有机肥7.5万t，沼液1.8万t。

Shangshui bio-natural gas project

Shangshui bio-nature gas plant was built by BioAll with Dry Plug-Flow AD technology in 2024. Annually, the plant treat 180,000~200,000 t corn stalks and produced 18.78 million Nm^3 bio-nature gas, 22,000 t food-grade CO_2, 75,000 t solid organic fertilizer, and 18,000 t biogas slurry.

商水生物天然气项目

工艺路线

以秸秆为主要原料，工艺包括秸秆贮存、横向塞流式干法发酵、净化提纯、CO_2生产、有机肥、微生物菌剂生产等单元。

Process route

Corn straw is used as raw material. The process includes storage and pretreatment, plug flow dry fermentation, purification, CO_2 production, organic fertilizer and microbial agent production.

∨ 项目工艺流程图
Process flow chart of the project

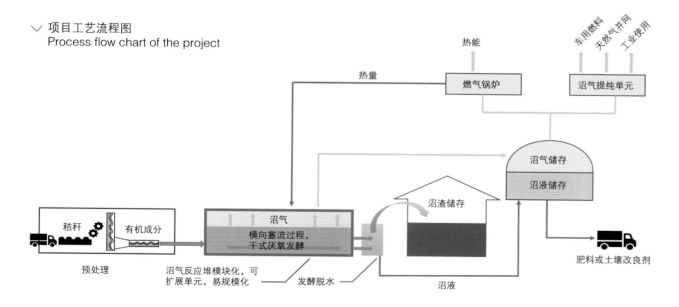

秸秆贮存单元

秸秆贮存区，占地 40 亩，贮存秸秆共 20 万 t。

Straw storage unit

Straw storage area, covering area of 40 mu, storing 200,000 t straw.

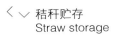
秸秆贮存
Straw storage

干发酵单元

项目建有 8 座干式塞流厌氧发酵反应器，采用连续进出料方式，通过皮带运输系统和螺旋搅拌器，每日将固含量 30% 左右的玉米秸秆 560 t 送入反应器内。厌氧反应器的顶部设有出气管道口，产生的沼气通过顶部管道输送至净化提纯单元。

Anaerobic gas production

The project includes eight containerized anaerobic reactors, utilizing a continuous feeding and discharging method. Through a belt conveyor system and spiral stirrers, 560 t corn stalks with a solid content of about 30% are sent into the reactors daily. The produced biogas is transported to the purification and refining unit through the top pipes.

∨ 干式塞流厌氧发酵反应器
Dry plug flow AD fermenter

商水生物天然气项目

∨ 沼气稳压气包
Biogas pressure stabilizer

净化提纯单元

项目选用变压变温吸附净化提纯工艺技术。甲烷纯度达 97% 以上。甲烷气干燥处理后，采用 CNG 压缩机压缩至 25 MPaG，充装入槽车对外销售。

Desulfurization and purification unit

Pressure swing adsorption (PSA) is applied in this project. Methane content is higher than 97%. After drying, the methane is compressed to 25 MPaG using a CNG compressor and filled into tankers for sales.

净化提纯单元 >
Desulfurization and purification unit

商水生物天然气项目

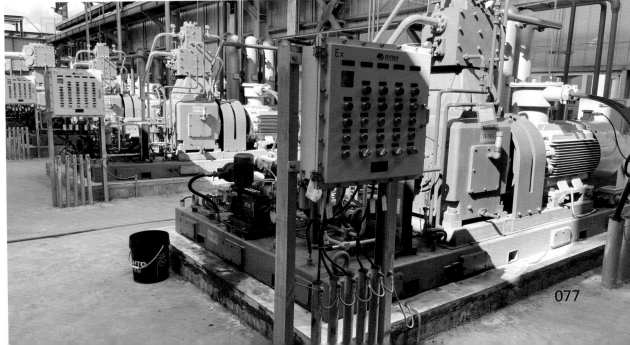

液态二氧化碳生产单元

变压吸附脱碳尾气二氧化碳，经深度脱硫，变温吸附脱水，进入液化器液化，得到食品级 CO_2 产品。

Liquid carbon dioxide production unit

The CO_2 desorbed from PSA decarbonization system undergoes deep desulfurization and dehydration. The liquid CO_2 then enters a purification tower to obtain food-grade CO_2.

液态二氧化碳生产单元 ＞
Liquid carbon dioxide production unit

商水生物天然气项目

有机肥生产单元

干发酵沼液产量少,储存罐共 3,000 m³,供后端制肥或回流利用。
沼渣固含量达 40%,项目年产固体生物有机肥料 7.5 万 t。

Organic fertilizer production unit

Low liquid digestate production. Storage tank 3,000m³ as organic fertilizer or for recirculation. Separated solid digestate has a solid content of 40%. 75,000 t bio-organic fertilizer are produced annually.

∨ 沼液储存罐
Liquid digestate storage tank

商水生物天然气项目

> 有机肥生产单元
> Organic fertilizer production unit

示范意义

项目将商水县 12% 的玉米秸秆离田高值化利用,供应了全县 55% 的天然气需求,有效破解当地有机废弃物对环境的污染问题,助力农业生产固碳减排,单个标准项目预计每年可减排 CO_2 35 万 t。

Demonstration significance

The project can utilize 12% of the corn straw in Shangshui County, fulfill 55% of the natural gas demand for the Shangshui County. GHG emission reduction could reach 350,000 t CO_2e annually.

商水生物天然气项目

射阳规模化生物天然气项目

该项目位于江苏省盐城市，由杭州能源环境工程有限公司建设，于 2021 年落地建成，占地面积 40 亩，总投资 8,900 万元，日处理农业废弃物 1,200 t，日产沼气 2 万 m³，提纯后可得生物天然气 1.2 万 m³，年减排 CO_2 5 万 t。

Large scale bio-natural gas project of Sheyang Jinyuan

Located in Yancheng City, Jiangsu Province, the project is constructed by Hangzhou Energy and Environment Engineering Co., LTD., and will be completed in 2021, covering an area of 40 mu with a total investment of 89 million yuan. It can treat 1,200 t agricultural waste per day, produce 20,000 m³ of biogas per day, obtain 12,000 m³ of biogas after purification, and reduce 50,000 t carbon dioxide per year.

射阳规模化生物天然气项目

工艺路线

采用"牛粪组合除砂及秸秆仿生水解预处理→高浓度高氨氮 CSTR 厌氧发酵→沼气生物脱硫（碱法）→沼气膜法提纯"的工艺路线，实现牛粪、猪粪及玉米秸秆的资源化综合利用。

Process rounte

The project is designed to adopt the process route of "combined desanding of cattle manure and straw bionic hydrolysis pretreatment → high concentration and high ammonia CSTR anaerobic fermentation → biogas biological desulfurization (alkali method) → biogas membrane purification".

> 工艺流程图
> Process flow chart

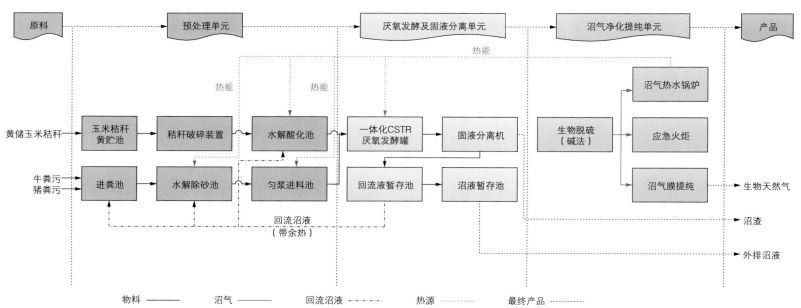

预处理单元

采用组合除砂工艺，由物料均质、除杂、除砂三部分组成，除砂率达 85% 以上。

采用秸秆"仿生"水解预处理工艺，停留时间约 2 d，厌氧发酵罐内的停留时间可缩短至 20~25 d。

Pretreatment unit

Pretreatment adopts homogenization, impurity removal and desanding process, ensuring the desanding rate is higher than 85%.

Adopting 2-day "bionic" hydrolysis pretreatment of the straw, the retention time of straw in the AD tank can be shortened to 20~25 d.

∨ 粪污预处理单元
Pretreatment unit of manure

∨ 秸秆预处理单元
Pretreatment unit of straw

厌氧发酵单元

4座一体化CSTR中温厌氧发酵罐，单罐有效容积5,000 m³，储气柜放置于罐顶。

Anaerobic digestion unit

4 mesophilic CSTR with each effective volume of 5,000 m³ and the gas storage cabinet is placed on the top.

∨ 一体化厌氧发酵罐
Integrated anaerobic digestion tanks

固液分离单元

螺旋挤压机对消化液进行高效分离，产生的沼液与沼渣进一步加工成生物有机肥与牛床垫料。

Solid-liquid separation unit

The screw extruder separates the digestate efficiently. Organic fertilizer and cattle bedding material are produced.

∨ 固液分离单元
Solid-liquid separation unit

沼气脱硫单元

利用湿法脱硫系统对原生沼气进行脱硫处理,降低对后端设备的腐蚀性。

Desulfurization unit

The wet desulfurization system desulphurizes the primary biogas.

∨ 湿法脱硫单元
Desulfurization unit

沼气提纯单元

膜法提纯技术对净化后的沼气进行脱碳处理。

Purification unit

The purified biogas is decarbonized through membrane purification technology.

∨ 提纯单元
Purification unit

示范意义

立足于3,500亩种植基地,以沼气工程为纽带,实现了"青贮饲料—牛粪—无害化处理—沼液还田—青贮饲料和水稻"的循环农业模式,解决了大型养殖企业的粪污处理问题,同时生产清洁能源与有机肥料,成为射阳农业的"新名片"。

Demonstration significance

The project realizes the cyclic production of "silage - harmless treatment to cow dung - biogas slurry returned to the farmland - silage and rice planting". It converts agricultural waste into clean energy and organic fertilizer, becomes a new business card of Sheyang agriculture.

射阳规模化生物天然气项目

大丰畜禽粪污集中处理综合利用项目

该项目位于江苏省盐城市,由杭州能源环境工程有限公司建设,于2015年投运。占地60亩,一期投资7,000万元,采用"原料预处理 + 中温厌氧发酵 + 三沼综合利用"技术路线,可日处理鸡粪600 t、日产沼气20,000 m³、生物天然气12,000 m³、沼渣肥15 t,年减排温室气体50,000 t CO_2e。

Animal manure centralized treatment and comprehensive utilization project in Dafeng County, Jiangsu Province

The project is located in Yancheng City, Jiangsu Province, constructed by HEEE in 2015. It can treat 600 t chicken manure per day. 12,000 m³ bio-natural gas and 15 t organic fertilizer are produced per day. Annual GHG reduction is 50,000 t CO_2e.

大丰畜禽粪污集中处理综合利用项目

工艺路线

以鸡粪为主要原料,工艺包括除砂除杂、全混式厌氧发酵、净化提纯与并网、有机肥生产等单元。

Process route

Chicken manure is used as raw material. The process includes desanding, AD, purification and injection, organic fertilizer production.

∨ 项目工艺流程图
Process flow chart of the project

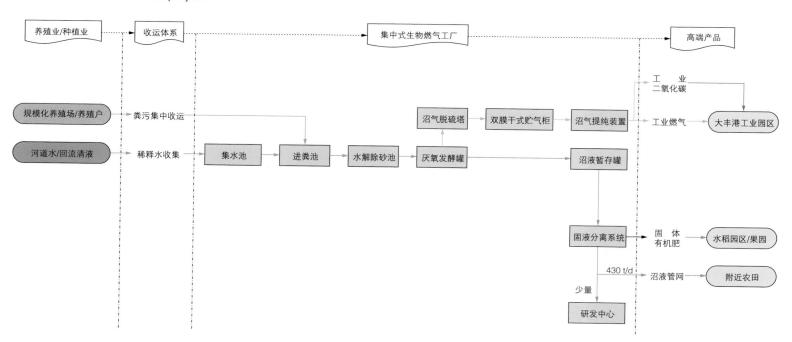

预处理单元

中温水解，实现粪砂分离。专用于鸡粪除砂，效率可达到 80%~90%。

Pretreatment unit

Mesophilic hydrolysis realizes sand separation of manure with removing rate of 80%~90%.

∨ 螺旋除砂机
Screw desander

厌氧发酵单元

4 座 3,500 m^3 厌氧发酵罐，采用 CSTR 工艺，中温厌氧发酵，容积产气率可高达 1.5 $m^3/m^3 \cdot d$。

Anaerobic digestion unit

Four CSTR tanks of 3,500 m^3 are adopted under mesophilic condition. The volumetric gas production rate can reach 1.5 $m^3/m^3 \cdot d$.

∨ 厌氧发酵罐
Anaerobic fermentation tanks

沼气存储单元

双膜干式独立气柜。

Biogas storage unit

Double membrane dry independent gas cabinet.

∨ 气柜
Biogas storage cabinet

沼气净化单元

湿法脱硫系统，能将原生沼气中的 H_2S 降至 0.01% 以内。

Biogas desulfurization unit

H_2S is reduced to less than 0.01% through wet desulfurization system.

脱硫系统
Desulfurization system

沼气提纯单元

采用 PSA 变压吸附提纯技术，沼气转化率 >92%，生物天然气纯度 >98%。

Biogas purification unit

PSA (Pressure Swing Adsorption) technology is applied. Biogas conversion rate >92% and bio-natural gas purity >98%.

∨ 提纯单元
Biogas purification unit

示范意义

项目形成"集中处理+生物天然气+商品有机肥+生态农业"的商业模式,将促进大丰生态高效农业快速发展,推进全区农业走循环路、环保路、生态绿色路,为沿海养殖集中地区提供可推广、可复制的畜禽养殖废弃物综合处理利用创新模式。

Demonstration significance

The project formed a business model of "comprehensive collection + commodity gas + commodity organic fertilizer + ecological agriculture", it promotes the region's agriculture to take the road of recycling, environmental protection, ecological green. Also, the project provides an innovative model of comprehensive treatment and utilization of animal manure that can be popularized and replicated for the coastal farming areas.

大丰畜禽粪污集中处理综合利用项目

镇江市餐厨废弃物及生活污泥协同处理项目

该项目由镇江市水业总公司建设，建于镇江市京口污水处理厂，占地 45 亩，投资 1.59 亿元，一期建设规模为 260 t/d，其中餐厨废弃物 140 t/d（含废弃油脂 20 t/d），生活污泥 120 t/d。采用"餐厨预处理 + 污泥热水解 + 污泥与餐厨高含固 / 协同厌氧消化 + 沼渣深度脱水干化土地利用 + 沼气净化提纯制天然气"的工艺路线。该项目年产沼气约 360 万 m³，年减排 CO_2e 48,606.3 t。

Zhenjiang food waste and domestic sludge collaborative treatment plant

Zhenjiang food waste and domestic sludge collaborative treatment plant was built by Zhenjiang City water industry Corporation, covers an area of 45 mu. 260 t/d of treatment capacity, including 140 t/d of kitchen waste (including waste oil 20 t/d), domestic sludge 120 t/d is designed for the first phase with construction investment of 159 million yuan. Co-anaerobic digestion of high solid content of thermal hydrolyzed sewage sludge and food waste+ deep dehydration and drying of solid biogas digestate for land utilization + bio-nature gas production and grid injection is used. 3.6 million m³ biogas is produced annually, which can reduce the CO_2e emission 48,606.3 t.

镇江市餐厨废弃物及生活污泥协同处理项目

镇江市餐厨废弃物及生活污泥协同处理项目工程全景图
Zhenjiang food waste and domestic sludge collaborative treatment project view

工艺路线

沼气除自用为热水解增温外,剩余部分经提纯、加压后直接并入市政燃气管网;毛油用于提炼生物柴油;沼渣经高干脱水后用于土壤改良、园林绿化等;沼液提纯后作为液态肥,剩余废水处理后达标排放。

Process route

The produced biogas is purified and pressurized. The produced bio-nature gas is directly incorporated into the municipal nature gas pipeline. The crude oil produced will be refined into biodiesel.
The digestate can be used for soil improvement and landscaping. The remaining waste water is discharged into sewage treatment plant.

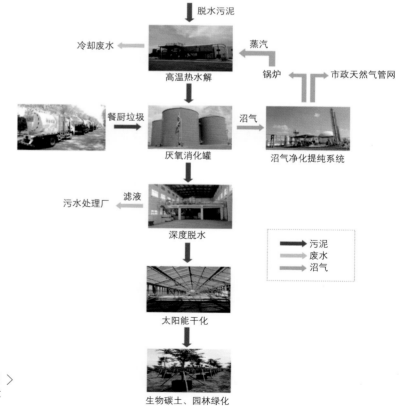

项目工艺流程图 >
Project process route chart

原料收集与预处理

采用特制运输车进行餐厨垃圾的收集、分拣、制浆。

Collection and pretreatment

Tailored transporters are used for collection, sorting and pulping of food waste.

特制运输车
Tailored transporter

高温热水解单元

处理规模 260 t/d,其中餐厨废弃物 140 t/d(餐厨垃圾 120 t/d、油脂 20 t/d),生活污泥 120 t/d,生活污泥脱水至含水率 80% 后进入污泥料仓。

High temperature thermal hydrolysis unit

Designed treatment capacity is 260 t/d, including 140 t/d of food waste (120 t/d of food waste, 20 t/d of grease) and 120 t/d of domestic sludge. Sludge with moisture content of 80% is thermal hydrolyzed.

∨ 高温热水解
High temperature thermal hydrolysis

镇江市餐厨废弃物及生活污泥协同处理项目

高含固厌氧发酵单元

建设 4 座厌氧消化罐，直径 16 m，单体有效容积为 2,700 m³。

High solid anaerobic digestion unit

4 AD tanks have been built, with a diameter of 16 m, and the effective volume is 2,700 m³ of each.

∨ 高含固厌氧消化系统厌氧发酵区域
High solid content anaerobic digestion system

沼气贮存单元

双膜沼气柜，双层膜球型结构，直径为 16 m，有效容积为 2,000 m³。

Biogas storage unit

Double-film biogas storage cabinet, with a diameter of 16 m and an effective volume of 2,000 m³.

∨ 沼气贮存系统
Biogas storage system

沼气净化提纯单元

沼气净化工艺为干法脱硫，沼气脱碳采用胺洗法，沼气脱水采用变温吸附（TSA）法。满足天然气二类标准要求。

Biogas desulfurization and purification unit

Dry desulfurization is applied. Biogas decarbonization is carried out by amine washing. Dehydration is performed by variable temperature adsorption (TSA). Bio-natural gas meets the second class stand of nature gas.

沼气净化提纯单元 ＞
Biogas desulfurization and purification unit

示范意义

形成了污泥与城市有机质协同厌氧消化 – 土地利用的"镇江模式"。在镇江二期、九江、荆门、六安等地实现推广应用,起到良好的示范引领作用,具有良好的社会、经济、环境效益。

Demonstration significance

"Zhenjiang model" of collaborative anaerobic digestion of sludge and urban organic matter - land use is formed. This model is promoted in Zhenjiang Phase Ⅱ, Jiujiang, Jingmen and Liu'an, which plays a good demonstration and leading role, with excellent social, economic and environmental benefits.

∨ 沼渣太阳能干化工艺
Digestate solar drying process

镇江市餐厨废弃物及生活污泥协同处理项目

"移动森林"苗圃示范基地
"Mobile Forest" Nursery Demonstration Base

佛山可再生能源（沼气）制氢加氢母站项目

该项目由瀚蓝（佛山）新能源运营有限公司负责建设，位于佛山南海区狮山镇南海燃气发展有限公司狮山 LNG 储配站内。项目建成后沼气制氢每小时制氢量为 3,000 m³，日产氢气约 6 t，年产氢气约 2,160 t。结合氢能工业、交通等领域的应用，该项目对传统能源的替代效应，预计每年可减少 CO_2e 排放近 100 万 t。

Grandblue Foshan renewable energy (biogas) primary hydrogen production and hydrogenation station project

The project is located at Shishan Town, Nanhai District, Foshan. 3,000 m³ per hour hydrogen can be produced through the biogas-to-hydrogen process, which means hydrogen production of about 6 t per day and around 2,160 t per year. Annually, anticipated GHG emission reduction could reach 1 million t CO_2e.

佛山可再生能源（沼气）制氢加氢母站项目

工艺路线

采用"湿法脱硫、PSA 脱碳、压缩、水蒸气重整、变压吸附提纯、压缩充装"等工艺单元。

Process route

Six primary units: wet desulfurization, PSA decarbonization, compression, steam reforming, pressure swing adsorption (PSA) purification, and compression filling.

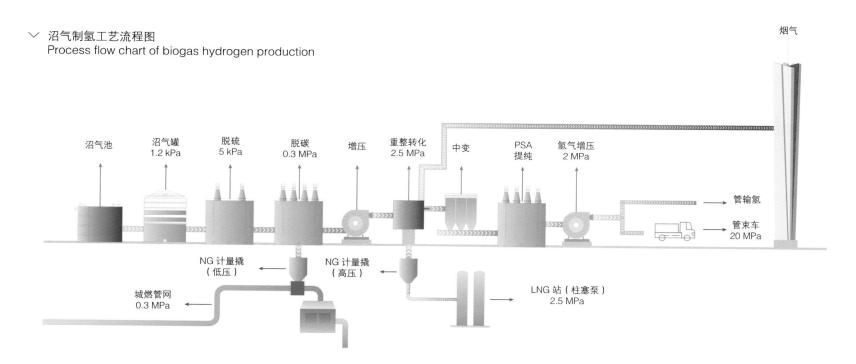

∨ 沼气制氢工艺流程图
Process flow chart of biogas hydrogen production

厌氧发酵单元

以餐厨垃圾厌氧发酵处置工艺中副产的沼气为原料，取代原有沼气发电工艺，生产附加值更高的氢气。

Anerobic digestion unit

Biogas generated from food waste is used to produce hydrogen, instead of power generation.

餐厨垃圾厌氧发酵处置工艺项目
Illustration of food waste anerobic fermentation and disposal process diagram

重整制氢单元

变压吸附获得 CH_4，经增压后去转化制氢工段；
CH_4 与水蒸气发生重整反应生成富氢气体。

Methane reforming unit

The methane is obtained by pressure swing adsorption.
CH_4 and vapor are reformed to produce hydrogen.

〈 转化炉
Reformer

加氢单元

产品氢气一部分增压外运,一部分通过管道输送到加氢站。

Hydrogen application unit

A portion of the produced hydrogen is pressurized and outbound transport. Another portion is conveyed to the hydrogen refueling station via pipelines

加氢站
Hydrogen filling station

示范意义

作为全国首个大规模餐厨垃圾沼气制氢加氢一体化项目，实现氢气本地化生产，有效降低制氢和运输成本。可以依托加氢站进行外销业务，形成完整的氢气"制、加、用"一体化模式。

Demonstration significance

As the first large-scale integrated project of hydrogen production and hydrogenation from kitchen waste biogas in China, it realizes localized production of hydrogen and effectively reduces the cost of hydrogen production and transportation. Based on hydrogen refueling stations, the project has created a comprehensive model for hydrogen "production, refueling, and utilization."

佛山可再生能源（沼气）制氢加氢母站项目

大湾区餐厨废弃物制备生物天然气示范项目

大湾区餐厨垃圾有机质含量高、规模大、天然气需求量大、价格高，为餐厨垃圾制备生物天然气项目的落地提供了良好的基础条件，目前形成了"收集分选＋精分制浆＋油脂回收＋厌氧消化＋脱硫提纯＋并网出售"的典型工艺模式。

Demonstration of bio-natural gas production from food/kitchen waste in the Greater Bay

The large scale food/kitchen waste and good price of natural gas in the Greater Bay provide good basis for the implementation of the bio-natural gas project. The typical mode of "collection and sorting + refined pulping + oil recovery + anaerobic digestion + desulfurization purification + grid sales" has been formed.

大湾区餐厨废弃物制备生物天然气示范项目

广州南沙餐厨沼气净化提纯制备生物天然气项目

该项目由山东红枫环境能源有限公司设计、承建，2022年投产。最高日产沼气 36,000 Nm³，最高日产生物天然气 21,600 Nm³，年减排温室气体约 15 万 t。

Bio-natural gas production project in Nansha, Guangzhou

The project is built by Shandong Hongfeng Environmental Energy Co., Ltd. and put into operation in 2022. Up to 36,000 Nm³ biogas and 21,600 Nm³ bio-natural gas are produced daily. The annual GHGs emission reduction is about 150,000 t.

大湾区餐厨废弃物制备生物天然气示范项目

预处理单元

餐厨垃圾采用"物料接收 + 大物质分拣 + 精分制浆 + 除砂除杂 + 油脂回收和提纯"为主的工艺路线。

厨余垃圾采用"物料接收 + 粗破碎 + 磁选 + 筛分 + 有机质破碎分离 + 挤压脱水 + 除砂"为主的工艺路线。

Pretreatment unit

The main process route of food waste pretreatment is "material receiving+sorting+fine pulping+sand and impurity removal+oil recovery and purification".

The main process route of kitchen waste pretreatment is "material receiving + coarse crushing+magnetic separation + screening + organic matter crushing and separation + extrusion dehydration + sand removal".

餐厨垃圾预处理单元
Food waste pretreatment unit

厨余垃圾预处理单元
Kitchen waste pretreatment unit

厌氧发酵单元

三座 6,000 m³ 厌氧消化罐，系统有机质降解率 ≥ 85%，罐体使用寿命 ≥ 25 年，清罐周期 ≥ 10 年。单位进场原生餐饮垃圾沼气产量 77 m³/t 以上。

Anaerobic digestion unit

Three 6,000 m³ AD tanks are arranged. The degradation rate of organic matter is ≥85%, the service life of the tanks is ≥25 years, and the cleaning cycle is ≥10 years. The biogas production of the original food waste is more than 77 m³/t.

脱硫单元

湿式催化氧化法粗脱硫 + 干法精脱硫工艺,粗脱硫进气、出气 H_2S 浓度分别为 2% 和 0.005%。

Desulfurization unit

Wet catalytic oxidation process of coarse desulfurization and dry fine desulfurization is used. The coarse desulfurization decrease H_2S concentration from 2% to 0.005%.

脱硫塔 >
Thionizer

提纯单元

沼气压缩 + 深度脱杂 + 变压吸附（VPSA）提纯 + 调压输配工艺，产品气甲烷浓度 ≥ 95%。

Purification unit

Through biogas compression+deep impurity removal+pressure swing adsorption (VPSA) + pressure regulation transmission and distribution process, the methane concentration of bio-natural gas is ≥95%.

∨ VPSA 提纯装备
VPSA purification equipment

东莞麻浦餐厨沼气净化提纯制备生物天然气项目

该项目由山东红枫环境能源有限公司建设，2019 年投产。最高日产沼气 30,000 Nm^3，最高日产生物天然气 20,000 Nm^3，年减排温室气体约 14 万 t。

Bio-natural gas project in Mapu, Dongguan

The project was built by Shandong Hongfeng Environmental Energy Co., Ltd. and put into operation in 2019. Up to 30,000 Nm^3 biogas and 20,000 Nm^3 bio-natural gas are produced daily. The annual GHGs emission reduction is about 140,000 t.

脱硫单元

湿式催化氧化法粗脱硫＋干法精脱硫工艺，粗脱硫进气、出气硫化氢浓度分别为 0.4%~0.5% 和 0.005%。

Desulfurization unit

Wet catalytic oxidation process of coarse desulfurization and dry fine desulfurization is used. The coarse desulfurization decrease H_2S concentration from 0.4%~0.5% to 0.005%.

提纯单元

沼气压缩＋深度脱杂＋膜分离提纯＋调压输配工艺，产品气甲烷浓度 ≥ 95%。

Purification unit

Through biogas compression+deep impurity removal + membrane separation and purification+pressure regulation and transmission and distribution process, the methane concentration of bio-natural gas is ≥95%.

∨ 脱硫塔
Thionizer

∨ 膜分离提纯装备
Membrane separation and purification equipment

示范意义

作为大湾区甚至华南地区建设投产较早、规模较大的餐厨垃圾厌氧沼气制备生物天然气项目，产生的生物天然气作为低碳能源并入燃气管网送到千家万户以及工业企业，为大湾区有机废弃物无害化处理和能源化利用做出了积极探索和成功示范。

Demonstration significance

As the large-scale and earlier treatment project of food/kitchen waste in the Greater Bay and even South China, the bio-natural gas generated is incorporated into the gas pipe network as low-carbon energy and sent to thousands of households and industrial enterprises, which has made a successful demonstration of the harmless treatment and energy utilization of organic waste in the Greater Bay.

∨ 在线再生式深度脱杂
On-line regenerative deep impurity removal system

大湾区餐厨废弃物制备生物天然气示范项目

膜分离提纯装备
Membrane separation and purification equipment

喷油螺杆式压缩机
Screw compressor with oil injection

造纸废水沼气提纯制备生物天然气示范项目

针对造纸工业废水量大、沼气利用附加值不高的特点,相关企业开展了废水沼气净化提纯制备生物天然气的探索,形成了"造纸废水制沼 – 脱硫提纯 – 并网出售"的商业模式。

Demonstration of bio-natural gas production from wastewater of paper mill

Bio-natural gas production from paper mill wastewater has been demonstrated by relevant enterprise. The business model of "wastewater biogas - desulfurization and purification - grid sale" has been formed.

造纸废水沼气提纯制备生物天然气示范项目

东莞中堂纸厂沼气制备生物天然气项目

该项目由东莞能源集团投资建设，日利用沼气 70,000 Nm³，发电 30,000 kW·h，最高可产生物天然气 52,000 Nm³/d，并入天然气管网，减排温室气体约 36 万 t/a，一期于 2013 年建成投产，二期于 2017 年建成投产。

该项目由山东红枫环境能源有限公司设计、承建。

Bio-natural gas production project in Zhongtang paper mill, Dongguan

The project is invested and constructed by Dongguan Energy. 70,000 Nm³ biogas is processed and up to 52,000 Nm³ bio-natural gas is produced daily. It is integrated into the natural gas pipe network and reduces greenhouse gases by about 360,000 t per year.

The project is designed and constructed by Shandong Hongfeng Environment and Energy Co., LTD.

造纸废水沼气提纯制备生物天然气示范项目

脱硫单元

湿式催化氧化法粗脱硫+干法精脱硫工艺，粗脱硫进气、出气硫化氢浓度分别为1.5%~2%和0.005%。

Desulfurization unit

Catalytic oxidation process of coarse desulfurization and dry fine desulfurization is used. The coarse desulfurization decrease H_2S concentration from 1.5%~2% to 0.005%.

< 脱硫塔
Thoinizer

造纸废水沼气提纯制备生物天然气示范项目

提纯单元

沼气压缩 + 变压吸附（VPSA）提纯 + 调压输配工艺，产品气甲烷浓度 ≥ 95%。

Purification unit

Through biogas compression + pressure swing adsorption (VPSA) + pressure regulation transmission and distribution process, the methane concentration of bio-natural gas is ≥95%.

VPSA 提纯装备
VPSA purification equipment

调压计量输配

产品天然气经调压计量加臭后送入市政燃气管网。

Voltage regulating metering transmission and distribution

Bio-natural gas is injected into the nature gas pipeline network after pressure regulation, metering and odorization.

∨ 调压计量装置
Voltage regulating metering device

沼气发电自用

部分富裕沼气用来发电自用。

Biogas power generation for personal use

The surplus biogas is used for power generation.

∨ 沼气发电机
Biogas generator

山东华迈纸业沼气制备生物天然气项目

该项目由山东世纪阳光纸业集团有限公司下属公司投资建设，日利用沼气 30,000 Nm³/d，最高可产生物天然气 21,000 Nm³/d，减排温室气体约 14.5 万 t/a，于 2022 年建成投产。

该项目由山东红枫环境能源有限公司设计、承建。

Bio-natural gas production project in Huamai paper mill, Shandong

The project was invested and constructed by a subsidiary company of Shandong Century Sunshine Paper Group Co., LTD., which can use biogas 30,000 Nm³/d, produce bio-natural gas 21,000 Nm³/d maximum, reduce greenhouse gas about 145,000 t/a. It was completed and put into operation in 2022.

The project is designed and constructed by Shandong Hongfeng Environment and Energy Co., LTD.

造纸废水沼气提纯制备生物天然气示范项目

脱硫单元

碱式生物法粗脱硫 + 干法精脱硫工艺，粗脱硫进气、出气硫化氢浓度分别为 2.5%~3% 和 0.005%。

Desulfurization unit

Alkaline biological coarse desulfurization and dry fine desulfurization are used. The coarse desulfurization decrease H_2S concentration from 2.5%~3% to 0.005%.

〈 脱硫塔
Desulfurizer

提纯单元

沼气压缩 + 变压吸附（VPSA）提纯 + 调压输配工艺，产品气甲烷浓度 ≥ 95%。

Purification unit

Through biogas compression+pressure swing adsorption (VPSA) + pressure regulation transmission and distribution process, the methane concentration of bio-natural gas is ≥95%.

VPSA 提纯装备
VPSA purification equipment

沼气压缩单元

选用集装箱式喷油螺杆压缩机。

Biogas compression unit

Select container-type oil injection screw compressor.

∨ 沼气压缩机
Biogas compressor

产品气调压计量加臭输配单元

产品气经调压计量加臭后并入燃气管网。

Pressure regulating, metering and odorizing transmission and distribution unit of product gas

The qualified bio-natural gas is injected into the gas pipeline network after pressure regulation, metering and odorization.

∨ 调压计量加臭输配装置
Voltage-regulating metering odorizing transmission and distribution device

示范意义

作为造纸工业最早一批建设的两个生物天然气项目，为造纸工业沼气的高附加值利用做出了积极探索和示范。在国内造纸行业中，中堂纸厂沼气制备生物天然气项目是国内造纸行业规模较大，也是国内最早建成投产的工业沼气提纯制备生物天然气的项目之一，具有沼气制备生物天然气与沼气发电自用耦合两种利用方式，项目采用了"EPC+O"的运作模式，为项目十来年的稳健高水平运营提供了关键支撑。

Demonstration significance

As the earliest two biogas projects constructed in the paper industry, it has made active exploration and demonstration for the high value-added utilization of biogas in the paper industry. In particular, Zhongtang project is one of the largest bio-natural gas project in the domestic paper-making industry, and it is also one of the earliest industrial bio-natural gas projects completed and put into operation in China.

造纸废水沼气提纯制备生物天然气示范项目

致谢
Acknowledgements

在本书编纂和调研过程中，重庆市环卫集团有限公司、维尔利环保科技集团股份有限公司、北京盈和瑞环境科技有限公司、山高环能集团股份有限公司、青岛汇君环境能源工程有限公司、必奥新能源科技有限公司、瀚蓝（佛山）新能源运营有限公司、山东红枫环境能源有限公司、中国农业大学、同济大学等企业和高校为本书提供了案例支持。感谢肖鹏、况前、刘慧、周晓、刘淼、吴广民、赵春英、高建鹏、高强、邢帆、王晓、张岳、梁敬明、李梁、李斌哲、李玉蓉、陈鹏辉、吴兴国、王恩振、郑永辉、刘若彤、宫徽等为相关案例提供图文素材，使得本书的呈现更加翔实、严谨与美观。感谢中国沼气学会副秘书长高嵩对本书的支持与帮助。特别感谢张茂真先生为本书题写书名。